Contents

Acknowledgments

The Central Office of Information would like to thank the Department of Trade and Industry, OFTEL, BT, Mercury, and Cable & Wireless for their co-operation in compiling this book.

Cover Photograph Credit
BT

Telecommunications

London: H M S O

Researched and written by Reference Services, Central Office of Information.

© Crown copyright 1994
Applications for reproduction should be made to HMSO.
First published 1994

ISBN 0 11 701784 1

HMSO publications are available from:

HMSO Publications Centre
(Mail, fax and telephone orders only)
PO Box 276, London SW8 5DT
Telephone orders 071-873 9090
General enquiries 071-873 0011
(queuing system in operation for both numbers)
Fax orders 071-873 8200

HMSO Bookshops
49 High Holborn, London WC1V 6HB   071-873 0011
Fax 071-873 8200 (counter service only)
258 Broad Street, Birmingham B1 2HE   021-643 3740   Fax 021-643 6510
33 Wine Street, Bristol BS1 2BQ
0272 264306   Fax 0272 294515
9-21 Princess Street, Manchester M60 8AS   061-834 7201   Fax 061-833 0634
16 Arthur Street, Belfast BT1 4GD   0232 238451   Fax 0232 235401
71 Lothian Road, Edinburgh EH3 9AZ   031-228 4181   Fax 031-229 2734

HMSO's Accredited Agents
(see Yellow Pages)

and through good booksellers

Introduction

The telecommunications industry is one of the fastest growing sectors of the British[1] economy. In terms of revenue, the British market for telecommunications services, network and terminal equipment is worth over £25,000 million a year. During the 1980s the sector experienced a growth of over 6 per cent a year. This strong growth in telecommunications has been stimulated by a wide range of factors. The most significant include the changes in government policy towards telecommunications, economic growth, and the rapid development in telecommunications and information technology.

This book outlines the changing structure of the British telecommunications industry and describes the main services and products provided by BT, Mercury and other companies, including their international activities and future plans. A glossary (see pp. 82–85) provides explanations for the abbreviations and some of the technical terms used.

---

[1]The term 'Britain' is used informally in this book to mean the United Kingdom of Great Britain and Northern Ireland; 'Great Britain' comprises England, Wales and Scotland.

Development of Telecommunications

History of Telecommunications Pre-1980

The invention of electric telegraphy does not quite mark the beginning of telecommunications in Britain. A system of visual signalling, the Murray shutter telegraph, was used by the Admiralty in Whitehall to communicate with the fleet bases in the Revolutionary and Napoleonic Wars (1793–1815). This system later spread to the United States, where a line of signalling stations linked New York and Philadelphia until 1846.

British pioneer work in electric telegraphy was carried out by Francis Ronalds (1788–1873) in 1816 and by William Cooke (1806–79) and Charles Wheatstone (1802–75), who installed the first inland telegraph system along part of the newly-built Great Western Railway in 1838. The Post Office purchased all private telegraph companies and became responsible for the inland telegraph business in 1870.

The first telephone apparatus able to transmit sustained articulate speech was patented in the United States in 1876 by Scottish-born Alexander Graham Bell (1847–1922) and was based on the principle of electromagnetic induction, discovered by the British scientist Michael Faraday (1791–1867) in 1831. Britain's first telephone exchange was opened in London in 1879, and in the next few years many more telephone exchanges were installed, a few by the Post Office but most by a number of private sector companies. The main private companies combined in 1889 to form the National

Telephone Company. An agreement was signed in 1905 whereby the Post Office purchased the whole of the National Telephone Company's system when the latter's licence expired at the end of 1911. From 1912 until 1981 the Post Office had, with a few exceptions, a statutory monopoly of Britain's telephone business.

The first effective submarine telegraph cable was laid between England and France in 1851, and the first commercially used transatlantic telegraph cable was laid in 1866. International telephone calls were made possible following the laying of the first submarine telephone cable between England and France in 1891, although it was not until 1956 that the first transatlantic cable (TAT 1) was laid between Scotland and Newfoundland. In 1962 the space communications radio station at Goonhilly Downs, Cornwall, took a leading part in the first international experiments in communications using artificial satellites to provide telephone and television links with the United States.

Development of Telecommunications Post-1980

Major regulatory changes have occurred since 1980. Competition has been progressively introduced into all the major telecommunications markets in Britain, most notably in fixed network operations. British Telecom (trading as BT since 1991) has been privatised and government shares sold to the public in three stages; a second fixed network operator, Mercury Communications, was licensed in 1982 and began offering services in 1984; and from 1990 the market was fully opened up to competition. The liberalisation of customer apparatus supply has led to a marked growth in the use of terminal equipment such as second telephones, facsimile (or 'fax') machines and private branch exchanges (PBXs). A substantial number of licences were granted to support the strong demand for a wide range of different types of mobile communications. In line with the continued rapid globalisation of British industries, the use of international communications services has grown by an average of more than 10 per cent a year over the past five years.

The growth in the British telecommunications sector has been accompanied by significant advances in telecommunications technology. These have enabled a wider choice of services and systems to be provided to customers. Advances over recent years have greatly improved the quality and reliability of networks, and calls made over networks. Optical fibre cables, digital switching and transmission technology, and the use of satellites for intercontinental traffic are bringing substantial increases in capacity and quality of service. During and since the 1980s mobile communications technologies have been growing rapidly, with the introduction of cellular and personal communications networks, telepoint and mobile data. Progress continues towards a multi-function fully integrated digital telephone network (ISDN). This is capable of handling voice, data, text and images. Even as ISDN services are starting to be used by customers, the next generation technology, broadband, an even more powerful and flexible version of ISDN, is under development. Throughout Britain, cable television companies are rapidly constructing broadband cable networks with the aim, in the long term, of providing integrated entertainment and telecommunications services.

Status and General Trends

Status

Britain has one of the most developed telecommunications markets in the world. The introduction of strong market deregulation since the early 1980s, economic growth over much of that period, and the technological progress made over recent years have all contributed to a telecommunications market which is among the most advanced in the European Community (EC). The telecommunications network infrastructure in Britain is highly developed and modern. At the end of March 1992 there were 25.7 million directly connected main telephone lines, representing 44.7 lines per 100 of the population. This is one of the higher line penetrations in the EC. There are also over 2 million users of mobile communication services in Britain.

Over 75 per cent of main telephone lines in Britain are connected to the homes of residential customers and by 1992 some 88 per cent of households in Britain had a telephone, compared with 52 per cent in 1975. The remaining main telephone lines are used by business customers and are linked either directly or indirectly via private branch exchanges (PBXs)[2] and key systems.

In fixed network communications the use of advanced digital technology and optical fibre cabling has led to substantial increases

[2]PBXs are directly connected to the public network. They are used to manage incoming and outgoing calls to and from a business customer's premises when the customer has a large number of telephone users in one location. Key systems are essentially small PBXs.

in service quality, network capacity and the choice of services available. Over recent years the most significant growth in terrestrial network communications has been in:

—the availability and choice in apparatus which can be connected to the network, from simple telephone handsets through to private telecommunications switches or PBXs; and

—the use of fax and dedicated data services such as the packet switched data network (PSDN) and private leased lines for business communications.

Progressive liberalisation, in conjunction with technological developments, has also stimulated the number, type and quality of mobile and satellite services available. In terms of subscribers, the mobile communications market in Britain is one of the largest in the EC. Over the past ten years or so there has been notable growth in:

—the availability and quality of cellular network services; and

—the use of paging services.

The take-up of telecommunications in Britain and comparisons with the telecommunications infrastructures of other national markets are shown in Tables 1 and 2.

Trends

The growth of the telecommunications industry has been accompanied by significant deregulation of telecommunications markets and advances in technology. As a result, telecommunications companies are providing increasing quality and choice in the services, systems and equipment offered to their customers.

Regulation

The 1981 and 1984 Telecommunications Acts promoted the Government's policy of liberalising all aspects of telecommunications

Table 1: Status of Telecommunications in Britain

Services (all suppliers)	Total	Date
Total public telephone main lines installed	25.7 million	March 1992
residential	19.7 million	
business	6.0 million	
Public payphones installed	Around 120,000	March 1992
Telex connections	60,000–70,000	Mid 1992
Packet Switched Data Network direct connections	54,000	1991
ISDN direct connections		1991
basic rate access	3,000	
primary rate access	9,000	
Cellular telephone subscribers	1.4 million	End 1992
Paging subscribers	710,000	End 1992
Telepoint subscribers	5,500	End 1992
Public Access Mobile Radio subscribers	40,000	End 1992
Fax machines installed	1.2 million	End 1992
Private Branch Exchange lines installed	9 million	1991
Telephone handsets installed	35 million	1991
Public electronic mailboxes installed	150,000	Mid 1992
Videotex subscribers	150,000	1991

Sources: annual reports, company literature.

in Britain. Competition has been introduced into the running of networks, the provision of services over the networks, and the supply of apparatus for connection to them. All British telecommunications markets have been progressively deregulated since 1984, the most notable steps in the process including:

—the licensing of Mercury Communications as the second fixed-link competing operator; and

—the award of licences to a large number of independent operators to provide a diverse range of non-basic communications services such as cellular, telepoint, mobile data, personal communication networks (PCNs), cable television, and two-way satellite communications.

The key events in the liberalisation of telecommunications in Britain are outlined in Table 3 and are discussed in further detail in the Regulatory Environment chapter.

Table 2: Comparative National Telecommunications Statistics 1989

Country	Fax machines	Cellular telephone subscribers
	(000)	(000)
Britain	670	975
Belgium	60	31
Denmark	n.a.	124
France	580	169
Germany[a]	920	164
Irish Republic	40	14
Italy	203	66
Japan	4,300	490
Netherlands	180	56
Spain	360	30
United States	4,400	3,500

Sources: International Institute of Communications (IIC), World Bank.
[a] Former Federal Republic of Germany.
n.a. = not available.

Table 3: Key Telecommunications Events in Britain in the 1980s and 1990s

Year	Milestone
1981	British Telecommunications Act 1981. The postal and telecommunications businesses of the Post Office are separated and British Telecom (BT) established to run telecommunications activities. British Approvals Board for Telecommunications established.
1984	Telecommunications Act 1984. Mercury Communications licensed to compete with BT, thus establishing a duopoly in fixed network operations. 51 per cent of the Government's shares in BT sold to the public. Establishment of an independent regulatory authority, the Director General of Telecommunications, supported by the Office of Telecommunications.
1985	Two cellular network operators licensed: Racal Vodafone (now Vodafone) and Cellnet.
1988	Two national and 16 regional private mobile radio operators chosen during 1987–88. Four operators selected to run telepoint services. Seven specialised satellite services operators chosen to provide one-way point to multi-point national satellite services.
1989	Three personal communications networks selected. Liberalisation of resale within Britain.
1991	The White Paper, *Competition and Choice: Telecommunications Policy for the 1990s*, is published. The Government decides that the duopoly of BT and Mercury Communications should be ended. Liberalisation of international one-end resale. The Government sells a further 27 per cent of BT shares to the public, retaining 22 per cent.
1992	Liberalisation of international simple resale (ISR) for designated countries; first ISR licence issued.
1993	The bulk of the Government's remaining stake in BT is sold.

The most recent move towards full market liberalisation resulted from the 1991 White Paper, *Competition and Choice: Telecommunications Policy for the 1990s.* The main proposal of the document was that the duopoly policy, under which only BT and Mercury were permitted to operate fixed telecommunications networks, should be ended and that the provision of such services should be fully open to competition.

Britain's liberalised telecommunications regime completely satisfies EC objectives regarding the establishment of a common European market for telecommunications services and equipment supply. These objectives were defined following an EC Green Paper published in 1987 and have been subsequently supported by a range of individual measures, in the form of EC directives issued from 1988 onwards.

The 1987 Green Paper, concerning the development of a common market in telecommunications goods and services, formed part of the EC's more general policy of establishing the single market in Europe by the end of 1992.[3] The aim of the Green Paper was to initiate discussion on the future regulation of telecommunications within the EC and its 12 member states. It argued that greater liberalisation and competition in telecommunications were essential if the EC was to realise its full potential.

After extensive consultations the European Commission set out policy objectives and a timetable for their accomplishment by EC legislation. These included:

—the rapid and full liberalisation of the market for terminal equipment;

—the gradual opening of the services market to competition from 1989 onwards; and

[3]For more information on the single market, see *Britain in the European Community* (Aspects of Britain: HMSO, 1992).

—the gradual implementation of the general principle that tariffs should follow overall cost trends.

The Commission also proposed accompanying measures to promote competition. These included:

—the complete separation of the regulatory and operational functions of EC telecommunications administrations;

—the rapid introduction of full mutual recognition of equipment type approvals throughout the EC;

—the definition of details for open network provision (ONP) which calls for full EC harmonisation of telecommunications network interfaces, standards and tariffs; and

—the establishment of a European Technical Standards Institute (ETSI).

During the period 1988 to 1992, a range of EC directives have been issued concerning liberalisation of the telecommunications terminal equipment, value-added and data services markets. A framework has also been set out for the establishment of ONP with specific regulations regarding leased lines. The ETSI was set up in 1988 and a directive regarding the mutual recognition of terminal equipment throughout the EC has been issued.

In June 1993 the European Community agreed to require member states to open up their markets for telephone services. The date set for this is January 1998. This will apply to most EC member states, including Britain, although one country has a derogation so that it need not comply until January 2000 at the latest, and three other member states have until January 2003. The Government has welcomed this agreement. The Commission also intends to examine opening up markets for mobile and satellite services.

Technology

The three most significant basic technologies shaping the telecommunications environment of the future are microelectronics, fibre optics and software. Advances in microelectronics, a technology which enables more information and processes to be built within computer chips, mean that complex processes which were until recently uneconomic for commercial applications can now be incorporated in consumer telecommunications products such as mobile telephone handsets. Progress in fibre optics has resulted in large reductions in the cost of long distance telecommunications transmission and substantial increases in fixed network capacity. Developments in telecommunications software provide a means of controlling telecommunications networks and are replacing the less reliable electromechanical technology. From these basic technologies are emerging strong trends at the network systems level. These include:

—increasing use of digital systems;

—the convergence of voice, data, text and image technologies; and

—the demand for mobility.

The Government is seeking ways of fostering innovation in the capabilities of the new network systems such as digital technology, the integrated voice, data, text and image technologies, including ISDN and broadband services; new mobile communications technologies, including PCNs; pan-European facilities; high-definition television; and satellite services.

Some 64 per cent of customers' lines are currently connected to digital exchanges, compared with 10 per cent in 1988. The links between telephone exchanges in the national trunk network are now wholly digital and typically based on fibre optical cabling. The transition to digital technology in the fixed network is stimulating the development of new advanced technology services running over these networks and improvements in the overall quality of services provided. It is also enabling the introduction of advanced voice services such as the forwarding or barring of calls to certain destinations, flexible billing methods for customers, and advanced data and image communications services such as electronic mail (E-mail), electronic data interchange, video conferencing and videophones.

The most notable trends in mobile communications technology are, again, in the introduction of digital technology, and also the reduction in the size of radio cells, particularly in city areas. Cell reduction facilitates an increase in the capacity of a cellular network to accommodate more users, and the use of lighter cellular telephones. These trends are allowing existing cellular operators to increase the capacity, quality and choice of their services, such as pan-European capabilities and voice messaging services, as well as the introduction of PCNs.

Investment

During the 1960s and 1970s the focus of telecommunications investment in Britain was on the development of the basic telephone network infrastructure and increasing the telephone line penetration of the population.

A modern and efficient telecommunications network is now established in Britain and the focus of investment in the 1980s and 1990s is being directed towards expanding the number and type of telecommunications services which run over the basic network. The relative cost of launching these advanced services, compared with initial infrastructure development, is low and as a result the investment per head on telecommunications is falling in Britain, while the range of telecommunications services offered is increasing rapidly.

Regulatory Environment

Regulatory Climate

During and since the 1980s there have been major changes affecting the regulatory climate for telecommunications in Britain. The British Telecommunications Act 1981 and the Telecommunications Act 1984 provide the legislative basis for the Government's policy of liberalising all aspects of telecommunications. The main aim of government policy is to establish a competitive telecommunications industry which is more responsive to consumer demand. Competition has been introduced into each of the main aspects of telecommunications in Britain:

—the operation of fixed, mobile and satellite communications networks;

—the provision of services over the networks; and

—the supply of apparatus for network connection and operation.

Network Operation

The British Telecommunications Act 1981 established a new framework for the progressive liberalisation of telecommunications services and resulted in a separation of the Post Office's postal and telecommunications businesses. The telecommunications business was restructured to form a new corporation, British Telecommunications plc (trading as British Telecom until 1991 and as BT since). The Government also became the licensing authority for the telecommunications operators as well as becoming responsible for setting standards and approving equipment.

Following the Telecommunications Act 1984, a 'duopoly' policy for the provision of basic services over national fixed networks was adopted. Mercury Communications (referred to as Mercury) was licensed as the second fixed-link operator. The company would provide a wide range of telecommunications services in competition with BT. In order to provide Mercury with some security in the early stages of its development, given the substantial investment involved, and to give BT time to adjust to competition in the private sector, the Government stated that it did not intend to license any further fixed network operators before 1990.

The activities of Kingston Communications (Hull)[4] were, however, considered as a separate entity to those of BT and Mercury. Kingston Communications, together with the city council, is responsible for public telecommunications in the Kingston upon Hull area. Mercury is allowed to compete in this part of the country but BT is not.

Under the 1984 Act, an independent regulatory body for the telecommunications industry was established. The Office of Telecommunications (OFTEL) is a non-ministerial department headed by the Director General of Telecommunications. The main responsibilities of OFTEL are described in further detail on pp. 22–9.

Consistent with its more general policy towards wider share ownership, 51 per cent of the Government's stake in BT was sold to the public in 1984, with a further 27 per cent disposed of in 1991. The Government sold the bulk of its remaining holding in BT by means of a public offer in July 1993. All of the Government's shareholding in Cable & Wireless (the parent company of Mercury) was sold in three stages starting in 1981.

[4]The historical development of Kingston Communications is discussed in further detail under the section on **Network Communications**.

At the time of the 1984 Act the Government made clear that the duopoly restrictions on the provision of basic fixed services did not apply to mobile and specialised services such as satellite and telecommunications over cable television networks. These markets have been progressively liberalised since 1984.

Further market liberalisation policy introduced during the 1980s which had an impact on network operation included changes concerned with payphones and the interconnection of public and private networks. In 1987 BT's monopoly in the supply and operation of public telephones in public places ended when OFTEL permitted Mercury to provide a public callbox service. The market for payphones in semi-public places such as airports and hotels and on private premises was deregulated in 1988.

Provision of Network Services

The market for telecommunications services which are conveyed over the network was progressively opened during the 1980s. The value-added and basic data services, such as computer networks and information services, were the first network services markets to be opened.

Network services and the resale of capacity on leased services are authorised under the terms of the telecommunications services licence (TSL). This describes the parameters within which network services may be provided, and authorises all forms of resale within Britain and various forms of resale of capacity on international circuits.

Equipment Supply

One of the first steps taken in liberalising the telecommunications market in Britain was to give customers choice over the apparatus which they could connect to the network. This ranged from the basic telephone handset, through fax machines and computers, to more sophisticated equipment such as private branch exchanges.

Since 1981, private sector organisations have been able to supply, install and maintain most types of subscribers' equipment attached to the telecommunications network. BT's monopoly over the supply and maintenance of a customer's first piece of telecommunications equipment ended in 1985, and since 1986 maintenance of newly installed apparatus has been fully open to competition.

The British Approvals Board for Telecommunications (BABT) was established in 1981 to provide independent evaluation and approval of such privately provided equipment. Since 1981, approval has been granted to more than 13,000 different items of equipment.

Measures were introduced in 1987 enabling applicants to submit simple equipment, such as telephone handsets, PBXs and other equipment used at a customer's premises, for evaluation direct to test laboratories instead of through the BABT. These have speeded up approval arrangements. The green circle label indicating that a telephone is approved by the BABT guarantees that the instrument complies with existing standards and has been approved for connection to the network.

Large and multinational businesses, with potentially complex private networks, and employing telecommunications specialists, are responsible for planning their own private networks in accordance with the networking code of practice. The code specifies network design criteria before connection is made with the public network in order to provide satisfactory quality of transmission. All individual pieces of telecommunications equipment must still be approved.

1990 Duopoly Review

In November 1990 a major review of government telecommunications policy resulted in the publication in March 1991 of a White

Paper, *Competition and Choice: Telecommunications Policy for the 1990s*. The main proposal in the White Paper was that the 'duopoly policy', under which only BT and Mercury were permitted to operate fixed telecommunications networks, should be ended. The Government would consider sympathetically applications from companies wishing to run fixed telecommunications networks in competition with BT and Mercury.

Background to the Duopoly Review

The 1990 duopoly review followed on from two previous statements made by the Government. In November 1983, the Government decided not to license operators other than BT or Mercury to provide basic fixed telecommunications services, but that this arrangement would be reviewed after seven years. The seven years of duopoly allowed Mercury some security in the early stages of its development. BT was also given time to adjust to competition in the private sector.

In July 1986, in a report entitled *Financing the BBC*, the Peacock Committee recommended that national telecommunications operators such as BT and Mercury should be permitted to convey a full range of services including television. The Government decided that it would examine this recommendation further when the duopoly policy was reviewed.

The main aim of the Government's telecommunications policy is to ensure that consumers—both business and residential users—have the widest possible choice of high quality telecommunications services at the most competitive price. Telecommunications services are considered in the three major market sectors: local, national (also referred to as 'trunk') and international services.

Local Services

In order to achieve greater competition in the provision of local services the Government stated that it would consider favourably applications for new licences for local fixed links. Factors which would be taken into account by the Department of Trade and Industry (the government department with responsibility for licensing telecommunications services—see p. 29) when considering a licence application would include:

—the type of service proposed;

—the ability of the company to fulfil its business plans; and

—the extent to which effective local competition would be introduced.

Prior to the duopoly review, cable television operators were not permitted to offer voice telephony services over their networks, except as agents of BT or Mercury. In the White Paper, however, the Government reconsidered this position and made two important decisions:

—cable television operators would be able to provide telecommunications services in their own right; and

—BT and other telecommunications operators can convey but not provide entertainment services over their national networks, this decision to be reassessed at least ten years after the duopoly review.

The final recommendations affecting the provision of local telecommunications services concerned radio communications. The Government aims to promote increased flexibility for existing mobile operators and licensees in offering fixed network services. In particular, support is given for the use of:

—radio access technologies, so-called 'radio tails', to make the final link to the subscriber's premises; and

—telepoint-to-the-home services, which make use of cordless telephones (see p. 58).

BT, Mercury and any prospective network operator will not be permitted, however, to provide mobile or telepoint-to-the-home services under the terms of the main fixed operating licences. Mercury's involvement with Mercury One-2-One, the mobile PCN licensee, falls within a separate operating licence.

National Services

With regard to national network services, the Government stated that it would consider sympathetically applications from companies intending to run services in all or part of the trunk network market. For a trunk licence application, the Department of Trade and Industry takes into account the same criteria as for a local network application.

It was recognised that network interconnection is an important issue for a company seeking a trunk network licence. Few customers would be attracted to a new service which does not allow interconnection with all other British and international telephone subscribers. The duopoly review introduces a procedure which supports effective interconnection between the networks run by existing operators such as BT and Mercury, and those networks of prospective new trunk operators. Specifically, OFTEL has the power to set interconnection terms where agreement cannot be reached between operators. These terms cover the commercial aspects of interconnection and the compatibility network interface standards.

'Equal access' allows customers of a local network (the 'access' network) to send their long distance calls over the network of their chosen trunk operator. In the White Paper the Government states its commitment to the progressive introduction of equal access. Previously, a customer who subscribes to both BT and Mercury could choose to route a trunk call via Mercury's network by using a suitably programmed telephone. Otherwise, by default, the call would be automatically routed via the BT network. The gradual introduction of equal access would enable:

—increasing possibility of access to the networks of Mercury or prospective licensees as BT modernises its network. At present, 90 per cent of telephone customers are able to access Mercury's trunk service.

—a customer of Mercury or another prospective operator to dial a special access code instead of using specialised telecommunications equipment.

Many customers are able to use the Mercury network simply by dialling such a code before making their call; a phased programme is under way to extend this.

International Services

The Government concluded that it would be unlikely to license any new full facilities-based international operators in the short term. The commercial resale of capacity on international private circuits has, however, been liberalised where a customer's private circuit is connected to the public network at one end only under the terms of the TSL. The TSL also permits the resale of capacity on leased international circuits connected to PSTNs at both ends—international simple resale (ISR)—for data traffic to other members of the EC.

In addition, the Government is considering the possibility of liberalising other forms of ISR. The Government has announced its readiness to issue ISR licences for services between Britain and those countries whose regulatory regimes allow an equivalent freedom to provide such services in the reverse direction. Australia, Canada and Sweden have already been designated for this. ISR allows customers to bypass existing operators for the international portion of a call, thereby reducing the most expensive cost element.

Other Points

Other important points of the duopoly review were:

— the Government welcomed further development of the assets of major British companies with nationwide operations, such as British Rail and the electricity companies, for telecommunications purposes;

— the establishment of a new national numbering scheme; and

— the modification of operators' licences to allow for the introduction of number portability.

Since the duopoly review, major public telecommunications operator (PTO) licences have been granted to City Of London Telecommunications (COLT), Energis, Ionica, MFS, Scottish Hydro-Electric, ScottishPower Telecommunications and Torch Communications. COLT is focusing on business customers in the Greater London area. Energis, a subsidiary of the National Grid holding company, and Scottish Hydro-Electric are the first two electricity companies to expand their interests into telecommunications, using their infrastructure as a platform for installing new optical fibre networks. Subsequent licences have been also been issued to ScottishPower Telecommunications, a subsidiary of Scottish Power, and Torch Communications, a joint venture between Yorkshire Electricity and Kingston Communications (see p. 48). Ionica is installing a new national network using radio to provide the final connection to customers.

Other new licences have been granted for satellite operations, international resale services and various specialist applications.

Office of Telecommunications

OFTEL, a non-ministerial government department established under the Telecommunications Act 1984, is the independent regulatory body for the telecommunications industry. It is headed by the Director General of Telecommunications, among whose functions are to:

—ensure that licensees comply with the conditions of their licences;

—initiate the amendment of licence conditions by agreement or by reference to the Monopolies and Mergers Commission;

—promote effective competition in the telecommunications industry;

—provide advice to the President of the Board of Trade on telecommunications matters and the issue of all licences; and

—investigate complaints.

OFTEL and the BABT are responsible for the approval of apparatus. OFTEL also approves maintenance contractors and designates certain standards. The maintenance of registers for all licences, approved apparatus and contractors falls under OFTEL's responsibility.

OFTEL also liaises with telecommunications advisory committees. There are about 170 of these voluntary bodies in Britain representing the interests of consumers in their respective areas. In addition, there are separate national advisory committees representing England, Scotland, Wales and Northern Ireland, and two other advisory committees, one representing the interests of elderly and disabled people, and the other representing small businesses.

The structure of OFTEL consists of six branches, each reporting to the Director General and Deputy Director General of Telecommunications. The branches' responsibilities are as follows:

—Branch 1: PTO licences and competition;

—Branch 2: consumer and international affairs;

—Branch 3: liaison with advisory committees, information section and administration;

—Branch 4: legal advice to the Director General on telecommunications matters;

—Branch 5: technical advice to the Director General on telecommunications matters; and

—Branch 6: economic, statistical and accounting advice to the
Director General on telecommunications matters.

Price Control

Within the function of promoting effective competition, OFTEL
has special responsibility for price regulation of certain BT services. The conditions of the BT licence granted in 1984 included an
agreement whereby the average price of a collection (known as a
'basket') of BT services was to rise by no more than a specified
limit. This limit was originally set at three percentage points below
the increase in the Retail Prices Index (RPI). The basket of services
agreed between OFTEL and BT included such services as line
rental, local, national and international calls. Changes in the RPI
are used to measure the rate of inflation in Britain and the so-called
price cap on the basket is known as 'RPI–x'. Thus, the original
price cap was referred to as RPI–3. If the RPI falls, or rises by less
than the fixed percentage 'x', the company must lower the
weighted average prices for their services within the basket accord-
ingly.

In 1984 OFTEL and BT agreed that the RPI–3 price cap was
to last until 1989 and the limit and structure of the price capping
mechanism would be reviewed at appropriate periods by the
Director General following consultation throughout the industry.
For the periods 1989–91 and 1991–93 the price cap was RPI–4.5
and RPI–6.25 (with the addition of international call charges)
respectively. In June 1992, the Director General proposed that the
limit on average price increases for switched services should be

RPI–7.5 for the period 1993–97, with BT's connection charges also included in the basket. This was subsequently agreed by BT.

OFTEL monitors the full effects of all BT's price changes on customers and ensures that the company complies with the agreed price cap. In recent years, the relatively low rate of inflation has meant that BT has been required to reduce the aggregate price of the services covered by the price cap. The impact of price changes on residential customers is of particular importance to OFTEL. The telecommunications usage of a typical residential customer is represented by a residential services basket, known as the 'median residential bill'. By measuring the annual changes in this bill OFTEL can assess the effect of BT price changes on the residential customer and make sure that price increases remain below the rate of inflation. Table 4 sets out the movements in the median residential bill and the composition of this quarterly bill for the four years to 1991–92.

Table 4: Median Residential Bill

median bill (£) per quarter, residential services[a]

	1988–89	1989–90	1990–91	1991–92
Connection charges	1.74	1.38	1.38	1.38
Exchange line rental	13.95	15.35	17.13	18.46
Local calls	10.33	10.57	12.04	13.04
National calls[b]	7.18	8.35	8.25	9.42
International calls	1.36	2.01	1.66	1.36
Total (excluding DQ/IDQ)[c]	34.56	37.66	40.46	43.66
DQ/IDQ charges	0	0	1.09	0.36
Total	34.56	37.66	41.55	44.02

Source: OFTEL.

[a] Excluding value added tax.
[b] Calls to the Irish Republic are included with national calls.
[c] DQ and IDQ represent charges for national and international directory enquiries.

Consumer Affairs

The Director General also has a duty to promote the interests of consumers in respect of quality of telecommunications services. In addition to OFTEL's ongoing task of monitoring network reliability and the quality of basic services, other major areas of current activity include applying the Citizen's Charter[5] initiative to telecommunications, and the misuse of some premium rate services.

The quality of service provided by BT to its customers has improved markedly during the 1980s. Table 5 shows the reliability of BT's network during the period 1983–92. Mercury also provides a similar high standard of network reliability for its customers.

As part of the Citizen's Charter initiative, the Competition and Service (Utilities) Act was passed in 1992. Under the terms of the Act, the Director General has been given new powers in respect of voice telephony and fax services, the preparation of certain directories, directory information services and public call boxes. In certain circumstances OFTEL will set agreed standards of service to individual customers and require the payment of compensation if they are not met.

Premium rate services, typically offered by independent service providers using the BT, Mercury or Vodafone networks, enable callers to obtain information—such as sports results or weather forecasts—or entertainment over the telephone, paying an additional premium for the call. Since their introduction in the early 1990s there has been a rapid growth in the demand for and availability of such services. While the majority of services are of high quality and informative content, there have been problems of

[5]The Government launched its Citizen's Charter initiative in July 1991. Aimed at raising quality and increasing accountability in public services, the proposals also covered the privatised utilities.

Table 5: Network Reliability 1983–92

Period[a]	Local calls failed[b] per cent	STD calls failed per cent	Network only faults per line per year per cent
1983	2.7	5.9	n.a.
1984	2.4	5.1	n.a.
1985	2.0	4.4	n.a.
1986	1.7	4.1	0.20
1987 (March)	2.2	5.4	0.25
1987 (September)	2.2	4.3	0.20
1988 (March)	1.9	3.6	0.22
1988 (September)	1.7	3.5	0.19
1989 (March)	1.4	2.4	0.21
1989 (September)	1.1	1.7	0.17
1990 (March)	1.0	1.2	0.25
1990 (September)	0.6	0.7	0.16
1991 (March)	0.5	0.8	0.19
1991 (September)	0.3	0.5	0.16
1992 (March)	0.3	0.4	0.18
1992 (September)	0.2	0.3	0.16

Source: OFTEL.

[a] From 1987, figures relate to month cited.
[b] Calls which have failed due to defective equipment or congestion.
n.a. = not available.

service misuse. A code of practice to address the problems is administered by the Independent Committee for the Supervision of Standards of Telephone Information Services (ICSTIS). Service providers are obliged, through their contracts, to abide by the code. ICSTIS may recommend to the relevant operator that service providers be barred from the network if they breach the code. Live and 'adult' services must use a separate prefix agreed by ICSTIS. In addition, BT has introduced, free of charge, an optional call-barring facility to premium rate services for customers served by digital exchanges. In April 1992 the Director General withdrew recognition of the code for multiple caller 'chatline' services. These services can therefore no longer be provided over BT or Mercury lines. BT has announced that in 1994 it will be introducing an opt-in requirement for all 'adult' services, giving its customers further protection against misuse.

Numbering

OFTEL will manage a national telephone numbering scheme from early 1994. It has already begun to prepare for this role by publishing in June 1993 a consultative document on future numbering options after the proposed national code change, which was announced in July 1992 and will take place in April 1995. By inserting an extra digit '1' in front of the initial '0' in area codes, many new codes will become available. Among the proposals that OFTEL has put forward in its consultative document is that codes beginning '07' should be reserved for personal numbers which an individual could retain for life. This could enable different members of a family to choose their own ringing tones.

Emergency Calls

After a review of the handling of calls to the emergency services, OFTEL announced in August 1993 that BT and Mercury would continue to handle such calls. There had earlier been a proposal that an independent national bureau should be established to handle all emergency calls, but studies showed that this would have only limited benefits. A review group set up to examine the question decided, however, that to have emergency calls handled by many different network operators would be undesirable. BT and Mercury have therefore offered to provide emergency call services to other operators; the terms on which this is done can be referred to OFTEL for determination if agreement cannot be reached. Kingston Communications will continue to handle emergency calls within its licensed area.

Department of Trade and Industry

The Department of Trade and Industry (DTI) is a ministerial government department headed by the President of the Board of Trade, also known as the Secretary of State for Trade and Industry. The DTI, acting on behalf of the President of the Board of Trade, has responsibility for issuing telecommunications licences and verifying that the general trading conditions of such licences are feasible. The Director General of Telecommunications has responsibility for enforcing the conditions of licences. Two main categories of telecommunications licence may be granted by the DTI: individual licences and class licences.

Individual licences are granted to specific companies, for example to Mercury and to each cable television franchise holder. Only individual licensees may be granted 'code powers' or designated as a PTO, or both. Code powers enable operators, for example, to follow simplified planning procedures for terrestrial network construction and maintenance. Designation as a PTO confers additional rights and duties on the operator, such as an obligation to provide emergency telephone facilities for users.

Class licences are granted to a particular system class or company. The branch systems licence, for example, authorises the connection of a private user's network (including payphones) to the networks of public operators. Users may install their own wiring and take advantage of private lines leased from BT and Mercury.

The Telecommunications Act 1984 defines code powers, PTO status and class licences in further detail.

Network Communications

BT

History

From 1912 until 1981 the operation and provision of telecommunications systems were the sole responsibility of the Post Office. Under the British Telecommunications Act 1981, however, the telecommunications activities of the Post Office were split off and a new corporation, British Telecommunications, was formed.

In 1984, consistent with the Government's general policies of wider share ownership and the introduction of greater market competition, two major changes occurred which had a strong effect on British Telecom's development:

—British Telecom was reconstituted as a public limited company and 51 per cent of the Government's stake in the company was sold to the public; and

—a second telecommunications network operator, Mercury, was licensed to provide fixed services in direct competition with British Telecom.

Following the 1990 duopoly review (see p. 17), BT operates in a fully liberalised fixed telecommunications services market. In 1991 the Government sold a further 27 per cent of its shares in BT. The bulk of its remaining 22 per cent stake in the company was sold by a public offer in July 1993.

BT currently provides one the most developed networks in the EC. Its network and services include:

—20.1 million residential lines;

—6.0 million business lines;

—between 60,000 and 70,000 telex connections; and

—around 113,000 public payphones.

Financial Statistics

BT is one of the world's largest public telecommunications operators. In 1992–93 its turnover was over £13,200 million. In March 1993 the company employed about 170,000 people; staffing levels have declined in recent years as a result of voluntary redundancy offers. BT has fixed assets of some £16,400 million, which include transmission equipment and telephone exchanges, and a large fleet of vehicles. To meet the increasing demand for basic telephone and specialised services, large-scale investment continues at a high level.

Of the £13,200 million total revenue generated in 1992–93, £5,100 million was derived from national telephone calls and £1,800 million from international calls. £3,000 million was earned from other non-basic services such as the mobile communications, private network services, and Yellow Pages. Pre-tax operating profit amounted to £1,972 million.

At the end of 1992–93 there were about 2.3 million BT shareholders, including 1.9 million who held interim rights to the shares sold by the Government in 1991 and 1.1 million holders of the shares sold in 1984 or issued to employees. Some 60 per cent of total BT shares were held by institutions (mainly insurance companies and pension funds), and 18 per cent were held by individuals, including employees, with the rest, 22 per cent, held by the Government. This ownership pattern will have been considerably altered by the July 1993 government share offer.

Network Modernisation

Since BT was privatised in 1984, the company has invested more than £17,000 million. Of the £2,155 million invested during 1992–93, expenditure on transmission equipment amounted to about 39 per cent of the total (or £835 million), and £545 million was spent on telephone exchanges. The main focus of current investment is mainly on upgrading the local telephone network to digital technology and installing optical fibre cabling.

Currently, about 64 per cent of customer lines are connected to digital exchanges compared with 10 per cent in 1988. The local and trunk digital exchanges installed during the 1980s are mainly from the System X family and AXE 10 type (both fully digital and computerised). Digital technology, by which calls flow through the network in the form of electronic pulses, is more reliable than conventional electromechanical technology and needs less maintenance. Digital exchanges have advanced processors which permit a high traffic-handling capacity and make possible advanced voice and data services such as call waiting (see p. 35) or call barring. Overall, more than 80 per cent of customer lines are linked to either digital or other modern exchanges, enabling these customers to benefit from receiving itemised bills.

In 1992 BT had over 2 million km (1.25 million miles) of optical fibre cable laid in Britain, compared with 367,000 km (230,000 miles) in 1988. Optical fibres, invented in 1964, are tiny strands of ultra-pure glass which can carry voice, data, text and images in digital format. They can carry substantial amounts of digital information (including thousands of telephone calls) on pulses of light between 40 and 70 km (between 25 and 45 miles) before needing a repeater, an instrument which corrects and amplifies messages through a series of electronic devices and removes distortion before retransmitting them.

BT's research and development expenditure was £240 million in 1991–92, which represents around 2 per cent of the company's turnover. The BT research centre is based at Martlesham Heath in Suffolk. Research into advanced optical and radio network technology, which enables telecommunications companies to lower their costs and offer greater communication capabilities, is the highest priority. However, there is also increasing work in systems and software engineering to provide better managed and more so-called 'intelligent' national and global networks. Intelligent networks can support a wide range of high quality, sophisticated and integrated services such as videoconferencing, teleworking and videophone.

BT Services and Products

The combination of digital exchange switching and digital transmission techniques, using optical fibre cable and microwave radio links, is supporting an increasing range of services on the company's main network, as well as substantially improving the quality of telecommunications services for residential and business customers. Telephone network services make up by far the main part of BT's business, but there is also a rapidly increasing requirement for data communications, text and image services and private network services.

In addition to the full range of voice, data and value-added services, BT offers a well-established worldwide telex service and a wide selection of equipment for customers' premises. Such equipment includes fax and telex machines, private branch exchanges and key systems, telephone answering machines and basic telephones.

Telephone Network Services

Over 75 per cent of the total number of calls made over the main public telephone network are generated by basic voice services.

However, the use of other services offered over the main network is increasing strongly, most notably facsimile transmission. In 1992 there were over 1 million fax machines in Britain, compared with around 56,000 in 1986. There is a growing customer demand for advanced voice services (sometimes referred to as value-added voice services) such as 'LinkLine' whereby callers contact businesses from anywhere in Britain or from certain overseas countries at reduced or nil call rates, the cost being paid by the recipient of the call. In 1992, more than 5 million calls a week were generated from BT's LinkLine services.

BT's advanced voice services are available to those customers served by a digital exchange and include:

—automatic 'LinkLine' numbers that enable callers to contact organisations anywhere in Britain, either free (0800 numbers) or at local call rates (0345 numbers);

—call waiting, whereby a caller hears a message when the receiver's line is engaged and the receiver is alerted to the other call on the line by a short bleep;

—call diversion, which allows a user to divert calls from the telephone normally used to any other telephone, including mobile phones;

—three-way calling, which allows a conference between three people at different sites at the same time;

—an operator-handled 'Freefone' service, whereby a caller can telephone a business free of charge and without needing a complete telephone number;

—'Voicebank', which is a voice messaging service by which messages are stored in a computer and retrieved at any time by entering a personal code into a telephone; and

—a range of 'Callstream' numbers, which allows callers to obtain information, such as sports results or weather forecasts, by paying a premium call rate.

A wide selection of 'Callstream' premium rate services, with international capabilities, are offered by independent service providers using the BT network. The 'International 0800' service is currently available in the major European markets, North America and Australasia. 'VoiceCom International', the international version of Voicebank, was available in 11 countries at the end of 1992.

A range of general services are also available through the main telephone network, including:

—a free facility for emergency calls to the police, fire, ambulance, coastguard, lifeboat and air–sea rescue services;

—operator services for various chargeable services such as reverse charge and alarm calls, and general assistance with making calls;

—national and international directory enquiries; and

—messaging services such as telegrams and telemessages.

A number of cashless services are also available for use on BT's telephone network. 'Phonecard' public payphones use pre-paid magnetic encoded cards and an increasing number of these payphones also accept major credit cards. Phonecards are also being increasingly used by companies for promotional purposes as they can be designed to carry colourful advertising messages or corporate statements. Due to the originality and range of many of the cards available, phonecards make very attractive items for collectors, and a new hobby has grown up around them. The BT 'Chargecard' enables a customer to make a call from virtually any telephone in Britain and have the cost of the call billed to a nominated account. Internationally, customers can use the BT

Chargecard when telephoning Britain from over 120 countries or alternatively use BT's 'UK Direct' service to place reverse charge calls from over 62 countries.

Data calls can also be made over the main telephone network via two modems, one used by the caller and the other by the receiver. A modem is a small device which translates computer-generated data into telecommunications signals suitable for transmission on telephone networks, and then turns them back into computer code at the other end. This method of data transmission is suitable for low quantity, person-to-person communication; most high capacity data communication is handled by BT's dedicated data network and private network facilities.

Data Communications

Data communications is one of the fastest growing telecommunications markets in Britain. In recent years the market has been stimulated by the rapid growth in the use of information technology, especially personal computers. The use of information technology to support business activity is widespread in Britain, and companies and other organisations are now seeking to use data communication services to link and integrate their data terminals and databases.

BT's Global Network Services (GNS) is a portfolio of data networking services. There are direct data connections in 28 countries, with gateway connections to another 100 countries. It uses the internationally accepted X.25 protocol, which was formulated by the International Telecommunications and Telegraph Consultative Committee (CCITT), members of which include BT and most of the world's major telecommunications operators. The network spans 1,000 cities, with 4,500 nodes and over 50,000 customer access ports.

Frame Relay is a new protocol developed to meet needs for high-speed data applications. BT's Frame Relay service connects geographically separated local area networks of computers to BT's Global Data Network.

GNS also includes applications services: BT Messaging Services and electronic data interchange (EDI). Using the international X.400 protocol, BT Messaging Services connects private messaging networks and systems to each other and to other public and private X.400 services around the world. The service is also connected to the existing fax and telex networks, enabling the user to send the same text to an electronic mail user, as a telex or to a fax machine without having to retype it or use any extra equipment. EDI is the transfer of data by agreed message standards from one computer to another by electronic means. BT's service, EDI*Net, allows users to send their business documents in a standardised electronic format to an EDI*Net electronic 'mailbox'. These electronic documents are then validated, sorted and delivered to the trading partners' mailboxes for retrieval when convenient. The EDI*Net service matches communication speeds and protocols between dissimilar computers, and provides a means to overcome time-zone differences around the world.

BT also provides a range of on-line information services, collectively known as Electronic Information Services. These are available via a personal computer equipped with a modem and communications software. They give access to information from companies such as FT Profile, Infocheck, ICC and Jordans.

Private Network Services

The data communications requirements of businesses in Britain can also be met by a range of private circuits and services offered by BT. These include:

—KiloStream, a range of digital private circuits with transmission speeds of up to one megabit per second;

—MegaStream, available as digital private circuits with two and eight megabit per second transmission speeds; and

—SatStream, a range of private satellite circuits with transmission speeds of up to one megabit per second.

Analogue private circuits are also available from BT and these are mostly used by businesses and organisations for voice communications.

New Technology Services
In 1985 BT launched one of the world's first integrated services digital networks (ISDN). The ISDN gives customers a direct link to advanced services available on the digital trunk network. Either separately or fully integrated, ISDN can carry video images, graphics, colour faxes and complex bar-coded product information in addition to basic telephone and circuit-switched data. ISDN services offer substantially faster transmission than BT's other telephone and data services. There are two BT ISDN services:

—ISDN 2, which is a basic rate access service made up of two 64 kilobit per second channels for data transmission and one signalling channel; and

—ISDN 30, which is a primary rate access service with maximum data transmission rates of 2 megabits per second.

In effect, ISDN 30 channels have a higher data-handling capability and flexibility than ISDN 2. The BT ISDN is currently connected to ISDNs in 16 countries.

Broadband is a digital telecommunications technology, similar to ISDN, but with even higher data-handling capability and flexibility. Transmission speeds of more than eight megabits per second are possible over broadband networks. At present broadband networks which operate over fibre optic cable are mainly used in Britain by the cable television companies. BT, however, is testing a broadband network in Bishop's Stortford in Hertfordshire in order to research the possible interactive services of broadband, such as integrated telephony and entertainment services.

BT has launched a videophone for domestic use, the Relate 2000. This contains a small fold-down camera and screen unit to send and receive colour pictures; with this folded away it can be used as an ordinary telephone. The Relate 2000 uses conventional phone lines to transmit the images at a rate of 10 frames a second. It is compatible with other videophones built to conform to the M-VTS Marconi videotelephone standard.

The development of the digital accessibility of the BT network has stimulated a growth in other new technology services such as teleconferencing and teleworking. Teleconferencing, encompassing videoconferencing, audioconferencing and business television, enables meetings to take place even though the people involved are not present at the same site. Videoconferencing is provided between BT's nine public studios and 450 other studios worldwide. With audioconferencing, BT can link between 30 and 60 locations throughout Britain and internationally. Business television can link any number of sites anywhere in the world. Teleworking is defined as customers working full-time at a location away from the traditional central office or business premises. Typically, customers are based at home and use business communications services such as fax and personal computers.

Joint Venture

In June 1993 BT announced a £1,000 million joint venture with the North American telecommunications company MCI. The joint venture, which will be based in Washington DC, will be owned 75 per cent by BT and 25 per cent by MCI. At the same time BT announced a large investment in MCI. The joint venture will combine the enhanced voice and data services over an advanced global network to multinational companies throughout the world. The venture will provide BT with a significantly enhanced presence in the North American market. Under the agreement, MCI will be responsible for marketing jointly-branded products throughout North and South America and the Caribbean, while BT will be responsible for marketing in the rest of the world. The agreement is subject to a number of conditions, including approval by the European Commission. It is expected to be finalised in the first half of 1994 following completion of all regulatory and shareholder approvals.

Mercury

History

Mercury Communications (referred to as Mercury), which is owned 80 per cent by Cable & Wireless and 20 per cent by Bell Canada Enterprise, is licensed as a public telecommunications operator in Britain, in direct competition with BT. The licence has a duration of 25 years from 1984. In 1985 Mercury obtained full interconnection access to the BT network for both national and international services. By 1993 the company had won over 10 per cent by revenue of the British market for telecommunications services.

Mercury has one the fastest growing telecommunications networks in the EC. Mercury's customer base, market share and traffic levels all increased significantly during 1992–93. The company has:

—about 174,000 directly connected business lines, representing an increase of 22 per cent over 1991–92;

—about 511,000 indirectly connected[6] business lines, 28 per cent more than in 1991–92;

—about 336,000 indirectly connected residential lines, representing an increase of 68 per cent over 1991–92; and

—thousands of public payphones located in major street sites, airports and major rail stations.

In 1992–93 Mercury had a turnover of £1,199 million, representing an increase of 31 per cent over the previous year's result, and employed around 9,500 employees. Pre-tax operating profit amounted to £192 million.

Network Development

Mercury's main all-digital national optical fibre network is now over 5,690 km (3,560 miles) long, with a further 2,700 km (1,680 miles) of digital microwave links. Stretching from the north of Scotland to the south coast, the network serves over 100 cities and towns in Britain, while local cables in key cities and partnerships with local cable television companies enhance the availability of Mercury services. The official launch of Mercury in Northern Ireland in 1991 means that the national network now covers all of Britain's major conurbations, with 90 per cent of businesses able to use the services.

Since its launch in 1984, Mercury has invested over £1,200 million. Mercury plans to continue investing in developing its network over the next few years.

[6]Indirect connection enables customers to be connected to the Mercury network via the local lines of alternative operators.

Mercury Network Services

Mercury offers a full range of long-distance and international telecommunications services for both business and residential customers. In addition to voice and data transmission, the company supplies advanced messaging services, mobile telecommunications and a range of customer-premises equipment. Customers can have a direct link between their premises or use Mercury indirectly via the local link of an alternative operator.

Telephone Network Services

A comprehensive telephone service began in May 1986 with the launching of Mercury's switched service, Mercury 2100. This service is aimed at medium to large businesses and provides direct connection to the Mercury network by cable for customers on or near local schemes and by microwave link elsewhere. The Mercury 2100 Premier service also provides access to the service benefits of ISDN, such as call forwarding and call barring to specific destinations.

In 1987 the indirectly connected Mercury 2200 services became available. As a prerequisite of the Mercury 2200 service, users need to have a Mercury-compatible PBX or key system. Alternatively, 'Smart Box' or 'Access 2200 Box', electronic devices which link non-Mercury-compatible PBXs with the exchange line, can be used. In each case the 2200 services automatically route all outgoing calls in the most cost-effective way, sending the call via the Mercury or BT network as appropriate.

Mercury's residential service, Mercury 2300, is for residential users and small businesses regularly making long distance or international calls. To allow use of this service, users need to have Mercury-compatible equipment, namely telephones or facsimile machines with the Mercury blue button. From summer 1993

Mercury began the phased introduction of its Easy Access service, which enables users to dial 132 for service connection, thus removing the need for an authorisation code. Mercury 2300 now supports over 336,000 lines, and is the fastest growing Mercury service.

Other services include a digital centrex service, which was the first such service in the EC. This service provides business customers with all PBX facilities (such as inbound directly dialled calls and call forwarding) from specialised 'centrex' equipment located at Mercury's public exchanges, so that they do not have to purchase their own equipment.

A wide range of operator services are available via the Mercury network, including the following 24 hour services:

—a free facility for emergency calls to the police, fire, ambulance, coastguard, lifeboat and air–sea rescue services;

—national assistance for services such as fault reporting, general enquiries and alarm calls; and

—national and international directory enquiries.

As all Mercury customers are served by digital exchanges, they have access to a range of advanced voice services with national and international capabilities, including:

—Mercury Premium Rate Service (PRS), which gives independent companies the opportunity to generate revenue from a range of information services, such as sports results or horoscopes, with callers paying a premium call rate;

—automatic 'CallLink' numbers that enable callers to contact organisations anywhere in Britain, either free (0500 numbers) or at local call rates (0645 numbers); and

—Mercury 2700, which is a voice messaging service allowing users to send and receive spoken messages at any time by entering a personal code into a telephone.

Choice has opened up in many areas of telecommunications. A customer at a BT Phoneshop is faced with a wide selection of different telephones to choose from.

Mobile communications have expanded greatly, and new technical standards have widened the available choice. One of the newest networks launched is Mercury One-2-One, which uses digital PCN technology (see p.60).

Modern communications can enable people to work from home. Here one of BT's directory enquiry operators takes calls from her living room, thanks to an experimental teleworking scheme.

The BT/RNID Typetalk service makes it easier for deaf people to communicate with hearing people. These operators, who are linked to the deaf user by a text link, can type in messages dictated over an ordinary phone and read back to the hearing person the text messages typed by the deaf person.

Mercury Telecommunications

Microwave links play a major part in modern telecommunications networks. This is Mercury's distribution node at Stevenage.

A large proportion of international calls are handled by satellite, using ground stations such as this one at Brechin.

Very large amounts have been invested in Britain in the installation of advanced fibre optic networks. This engineer is at work splicing the cables.

Despite the introduction of satellite communications, advanced international cables are still being laid by vessels such as the Cable and Wireless cable-laying ship CS *Sir Eric Sharp*.

Data Communications

Mercury offers a comprehensive portfolio of value-added, data and messaging services. These include:

—two directly connected telex services, Mercury 7100 and Mercury 7125, the latter of which is a unique service providing telex and X.25 packet switching;

—an indirectly connected telex service, Mercury 7200, which enables customers to use Mercury tariffs by dialling a special access number; and

—MultiMessage, an integrated messaging system combining electronic mail with access to fax, telex and on-line databases.

National and international data transmission services are also provided over its public packet switched data network, including:

—two managed-data network services: Mercury 5000, accesses via a data link or dialled through the public switched telephone network, and 5000 CDN (for corporate data network), which provide digital links between computers throughout Britain using the company's X.25 packet switched network;

—Mercury 5000 International, which provides links to more than 160 public packet switched data networks in over 80 countries around the world;

—Mercury's Global Managed Data Services, which integrates the existing national and regional data network services group capabilities of Cable & Wireless subsidiaries around the world; and

—Switchband, a variable bandwidth data transmission service which allows companies to transmit high volumes of data at great speed without needing to have a large number of permanent private circuits.

Private Network Services

Mercury 1000 service provides dedicated permanent and private links between a customer's different sites and locations in Britain. Transmission of up to 8 megabits per second and international connections are available within the service.

The Mercury Global Virtual Private Network (GVPN) service allows customers to address their communications requirements using a mix of public and private voice and data services. The service can be used to replace dedicated private links which are not fully utilised by the customer or it can supplement them by providing additional capacity at peak times. Mercury offers a range of virtual private networks, both nationally and internationally, which offer high capacities without the capital costs of building a full private network. They are flexible and can be delivered to customers' or suppliers' sites. The GVPN is currently provided in Britain by Mercury on behalf of Cable & Wireless and is also available in North America and the Pacific Rim.

New Technology Services

Other new technology services offered by Mercury include video-conferencing, teleworking services, broadband services and specialised telecommunications systems designed to meet the requirements of specific customers, such as the Mercury Dealing System which is used by financial organisations in the City of London.

Cable & Wireless

The main business of Cable & Wireless is the provision and operation of public telecommunications services. It has operations in over 50 countries, including Australia, Hong Kong, Japan and the

United States, in most cases under franchises and licences granted by the governments concerned. It employs over 39,000 people, some 75 per cent of whom are based outside Europe. In 1992–93 turnover totalled £3,826 million, and pre-tax operating profit, excluding exceptional items, was £824 million. In Britain, Mercury Communications is largely owned by Cable & Wireless.

In recent years Cable & Wireless has been constructing and bringing into service a broadband digital network linking major world economic and financial centres in Europe, North America and the Pacific Rim region. Based on high-capacity optical fibre technology, the network is called the 'Global Digital Highway' (GDH—see p. 69).

The international services available include the Cable & Wireless Calling Card, launched in 1992, which allows customers to make national and international calls from telephones in any participating country and to have the cost of calls billed to a single account.

The company also operates the world's largest fleet of commercial ships and submersible vehicle systems for laying and maintaining submarine telecommunications cables.

Other Fixed Network Operators

Kingston Communications is the network operator for the Kingston upon Hull area of Britain. The company provides a full range of local voice services, with customers interconnecting with the Mercury network for national and international communication services.

The development of Kingston Communications has been somewhat unusual compared with the backgrounds of BT and Mercury. Unlike other areas in Britain, the local authority telecommunications venture in the Kingston upon Hull area was not fully absorbed by the Post Office in the early 1900s. In June 1984, the Post Office licence under which the Kingston venture had been operating was replaced by a licence defined by the Telecommunications Act 1984. In 1988 the undertaking was transferred into a separate company, Kingston Communications (Hull), which is wholly owned by Kingston upon Hull City Council. Kingston Communications, together with the city council, is responsible for the provision of basic public telecommunications in the area. The terms of Mercury's licence extend to this part of the country, but BT is not able to compete in this area. In September 1993 a PTO licence was granted to Torch Communications, a joint venture between Kingston Communications and Yorkshire Electricity. The new company plans to commence services in the Yorkshire and Pennines area.

Since the duopoly review, several new operators have received licences to provide telecommunications services over fixed networks (see p. 22).

Mobile Communications

Cellular Radio

Cellular radio is a mobile communications system which allows users to make or receive so-called two-way communications in different locations of their choice within the network coverage area. Users can be stationary or on the move during calls.

Cellular radio is by far the largest public mobile communications market in Britain. At the end of 1993 there were around 1.8 million subscribers to the two cellular radio networks. This represents a penetration of around 30 subscribers per 1,000 of the British population. The two cellular networks cover more than 90 per cent of the population. There is continued strong growth in the take-up of cellular radio services, with the number of cellular subscribers increasing by approximately 28 per cent during 1993.

Britain has the largest and one of the fastest-growing cellular markets in the EC, stimulated by a strong latent demand for mobile communications and progressive liberalisation of the mobile services sector. Germany and Italy have the second largest cellular markets, with each country having around 750,000 subscribers at the end of 1992.

The cellular radio system operates by dividing the country into a network of cell areas, each serviced by a base station which contains a computer, a radio transmitter and receiver and a directional antenna. The power of the radio equipment in the base station is fixed so that it covers only that particular cell. When a cellular radio handset is switched on, the computer at the nearest base station senses its presence within the cell and the handset is then able to transmit and receive messages from that position.

A user can also communicate on the move. As the vehicle moves from one cell to the next and it approaches the cell's outer limit, the strength of the cellular handset's radio signal falls below a predetermined level. At this point the base station computer instructs the next cell to identify the handset, releasing the old cell for use by another subscriber. The process by which the handset is transferred from one cell to the next is referred to as 'hand-off'. The Highway Code states that hand-held cellular phones should not be used by drivers on the move.

Cellular Network Operators

In 1983 the Government selected two companies to operate cellular radio networks in Britain: Racal Vodafone (now trading as Vodafone) and Telecom Securicor Cellular Radio (which trades as Cellnet). Since the two operators launched their networks in 1985 there has been rapid take-up of cellular services.

By the end of 1991 the two network operators had each invested over £700 million in establishing their networks, including the further investment required to extend network capacity. The operators each expect to spend at least £30 million a year over the period 1993–95 improving the quality and capabilities of their networks.

In addition to basic voice services, the cellular networks can provide a range of value-added facilities including cellular network, voice messaging, mobile data and private network services. Cellular network services, such as diverting a call to another number if there is no reply or if the cellular phone is already busy, can be preprogrammed into a user's handset. Voice messaging services enable incoming messages to be stored and retrieved later by the user.

Mobile data services enable communication to and from mobile data terminals such as facsimiles or personal computers. Cellular phones can be connected to users' fixed PBXs so that the cellular phone can be fully integrated into the office communication system.

Towards the end of 1992 Vodafone and Cellnet launched so-called 'low-user' packages for the consumer market. These services are based on an alternative set of tariffs aimed at those customers who make little use of the cellular network in comparison to major business users.

In 1985 Racal Vodafone was part of the British electronics company, Racal Electronics. During the late 1980s, the activities of Racal Electronics expanded and Racal Telecom was set up to run the various telecommunications ventures of the parent company. These ventures included Racal Vodafone. In 1991 Racal Telecom was fully separated from its parent company, Racal Electronics, and renamed the Vodafone Group. In terms of turnover, the operation of the cellular network is the largest activity of the Vodafone Group, but the company also provides paging, packet switched data and value-added network services. The Vodafone Group also has joint ventures for the manufacture of mobile telecommunications equipment, as well as mobile investments in Hong Kong, France, Sweden and Mexico.

Telecom Securicor Cellular Radio was formed in 1983 as a joint venture between BT and Securicor. The company trades as Cellnet. Ownership of the Cellnet Group, of which Cellnet is a subsidiary, is currently divided between BT (60 per cent) and Securicor (40 per cent). In 1992 the company structure of Cellnet was reorganised. The cellular network operations form the largest part of the group's business but other activities include service provision, billing systems and value-added services. The company launched its cellular network in January 1985 and by the end of

1991 had achieved a coverage of over 90 per cent of the British population.

Cellular Service Providers
Unlike the licences issued to other mobile communications operators in Britain, one of the conditions of a cellular network operator's licence is that the operator is not allowed to deal directly with end-users. Vodafone and Cellnet must therefore use third parties to provide cellular services to their customers. This restriction is being relaxed, however, as the network operators are being allowed to sell services direct to the end-user from January 1994. They will still be required to provide services to intermediaries if requested. The so-called cellular service providers currently have responsibility for:

—supply of cellular network airtime;

—equipment sales, rental, installation and maintenance;

—customer care and billing;

—credit control; and

—value-added services.

The majority of service providers have also appointed a group of dealers to sell and install equipment on their behalf.

Currently, there are approximately 30 cellular service providers in Britain, of which five have responsibility for around half of the cellular network subscribers. The main providers are:

—BTMC, a division of BT;

—Hutchison Cellular Services, a subsidiary of Hutchison Telecommunications (UK);

—Mercury Communications Mobile Services, a subsidiary of Mercury;

— Talkland; and

— Vodac, a subsidiary of Vodafone Group.

Cellular Equipment

There are three types of cellular radio handset, or cellular phone. These are the carphone or mobile, the transportable and hand-portable.

The carphone or mobile is fitted into the vehicle and receives its power supply directly from the vehicle's battery. The hands-free facility is voice activated and enables a caller to use the car-phone without having to hold the telephone handset. The carphone works anywhere within coverage of the network but the equipment can only be used in the vehicle.

The transportable phone is very similar to the carphone but is designed as an integral unit with a detachable battery pack so that it can be used outside the vehicle, although transportables generally have a lower power output when used outside the vehicle.

Handportables are the lightest cellular phones. They are battery powered by an integral or detachable battery which is rechargeable. Handportables can therefore be used inside or out-side a vehicle. The power output is lower than the carphone or transportable so handportables work best where network coverage is good, such as in cities or other built-up areas.

Cellular Services

To date, cellular radio technology in Britain has largely been based on the proprietary analogue technologies—Total Access Communications System (TACS) and Extended TACS (ETACS). TACS was the first basic cellular radio system used in Britain and utilised 300 TACS frequencies on the 900 MHz range of radio channels. As the number of subscribers grew rapidly the radio channels became fully utilised. In order to address the increasing problem of network congestion, an additional 280 channels, known as ETACS, were made available to each operator at the end of the 1980s.

The Special Mobile Group (Groupe Speciale Mobile or GSM) of the Conference of European Posts and Telecommunications Administrations (CEPT) has specified the technical standards necessary for the introduction of a pan-European digital cellular mobile communications system. At least 18 CEPT administrations or operating companies have signed a memorandum of understanding on co-ordinated action to implement this common, harmonised standard. Signatories (in Britain, Cellnet and Vodafone) have started to launch GSM services throughout Europe. By the mid-1990s, according to the signatories' plan, all European capitals, their airports and major traffic routes will be served by the new service.

During 1992 Vodafone launched its digital cellular service which forms part of the pan-European GSM service. The service was introduced in central London and the South East, and the network had achieved 50 per cent population coverage by the end of 1992. This new digital network will provide the foundation for a Micro Cellular Network (MCN), which is planned for introduction during 1993. The MCN is intended to broaden the market for mobile communications by introducing a cheaper service aimed at the small business and domestic markets.

Similarly, Cellnet aims to make its digital cellular GSM-compatible service available during 1993, focusing initially on providing services in London and major conurbations.

Radiopaging

The simplest mobile radio communications are provided by radiopagers. Radiopaging is also referred to as paging or wide-area paging. Pagers enable people to remain contactable cheaply and reliably while away from base. Pagers provide one-way communications—that is, a user is able to receive an incoming message but is not able to transmit a message from the pager. The recipient of the paging message must find a telephone to respond.

Britain has by far the highest number of subscribers in the EC. At the end of 1992 there were approximately 710,000 users of paging services, representing a penetration of around 12 pagers per 1,000 population, and over 97 per cent of the population can be reached by paging services. The annual growth is now steady after very strong take-up of services during the 1980s. The number of paging subscribers increased by around 5 per cent in 1992. Germany has the second highest number of paging users in the EC, with approximately 400,000 at the end of 1992, representing a penetration of around 5 pagers per 1,000 population.

During the 1980s the Government allocated a number of frequencies for paging services and eight companies were awarded licences for national operations. By the end of 1992, the industry had consolidated, however, and there are now five major operators providing paging services on a national basis. A major consolidation took place at the end of 1992 when Mercury Paging and Inter-City Paging merged to create the third largest paging operator in Britain, also called Mercury Paging.

The first entrant, BT Mobile Communications (BTMC), a division of BT, remains the largest operator in terms of the number of subscribers. The five major operators are:

—BTMC, a division of BT;

—Aircall, owned by Vodafone Group;

—Mercury Paging, jointly owned by Mercury Communications, and Motorola and Mtel of the United States;

—Vodapage, a wholly-owned subsidiary of the Vodafone Group; and

—Hutchison Paging UK, a division of Hutchison Tele-communications (UK).

There are three main types of pager available: so-called tone-only, numeric and alphanumeric. In effect, each of the pagers is a miniature radio receiver capable of handling different levels of information. The tone-only pager emits a distinctive bleep when the user's personal paging number is dialled. The numeric pager is able to receive a short numeric message from a caller, such as a contact telephone number, which appears on a liquid crystal display window on the pager. Similarly, the alphanumeric pager is capable of handling a more detailed message consisting of numbers and letters.

There are two main methods by which a paging service is provided to a customer. The first, 24-hour bureau answering, enables a caller to telephone a paging operator receptionist, who then relays the caller's message to the user by radio signal. The direct input facility allows calls to be made to all types of pager direct from a personal computer or terminal, or via the public data network or a leased line, without the need for a receptionist.

In addition to basic services, paging operators provide a range of value-added paging facilities, such as voice messaging services which enable incoming messages to be stored for access by the paging user, and Euromessage, whereby the British paging service is extended to other selected European countries. Euromessage is provided by certain European paging operators as an interim service before the pan-European paging service, European Radio Messaging Services (ERMES), is fully available throughout Europe. ERMES is a paging system being promoted by the

European Commission and the ETSI. The aim of ERMES is to have a European paging network with fully harmonised pan-European standards and network interfaces.

Private Mobile Radio

Private mobile radio, also referred to as PMR, is provided on a competitive basis in Britain. PMR services are aimed at closed-user groups of mobile communications services, such as taxi fleets, road haulage and transport distribution sectors.

At the end of 1992 the national PMR operator, National Band Three (see below), had approximately 30,000 subscribers to its network. In addition, there were a substantial number of individual private PMR users with local or regional operations.

In 1988, as a result of increasing congestion problems on permitted PMR channels and frequency restrictions, the Government released one-third of the frequency spectrum known as Band Three for commercial use, freeing hundreds of new channels. At the same time, two operators of national PMR networks were selected and by 1992, 13 operators of regional PMR services had also been awarded licences.

The two operators granted nationwide PMR licences were:

—Band Three Radio, in which stakes were held by Racal Telecom (now Vodafone Group), Philips, Securicor and Aircall; and

—GEC National One, which was wholly owned by GEC.

The national PMR market has recently consolidated. In 1992 Vodafone Group acquired 100 per cent of Band Three Radio, which was subsequently merged with GEC's National One to form the Vodafone Group/GEC joint venture National Band Three. In mid 1993, National Band Three was acquired by the American company Geotek.

In August 1993 the Government announced that the narrow band 5 KHz channel would be made available to PMR users to meet increasing demand. This makes use of spectrally-efficient technology that has been pioneered in Britain by Bristol and Bath Universities. The relevant technical standard has been cleared with the European Commission. The Government also announced that data transmissions would be allowed on speech-dominant shared PMR channels. Many PMR users share channels, but, in order to avoid interference, have not previously been allowed to transmit data on these shared channels. Tests have now shown that limited data facilities can be introduced on shared channels.

New Technology Services

Telepoint

Britain was the first country in the world to offer telepoint services. Similar services are available in the Netherlands, France, and some countries of South-East Asia.

Telepoint services are based on second generation cordless telephone (CT2) digital technology. A CT2 cordless telephone uses a radio link between the handset and a public telepoint base station. Typically, these light and portable telephones provide one-way communication. They are capable of transmitting, but generally not receiving, calls within a maximum of 200 m (660 ft) of a public base station.

More sophisticated telepoint equipment is available, which has paging facilities enabling the user to receive simple messages. Such a handset can also be operated in conjunction with a private base unit in the user's home connected to the fixed network. It then operates as a cordless telephone within range of a private or public telepoint base station.

Telepoint base stations are connected to the fixed telephone network by exchange lines, currently supplied by BT or Mercury, and thus the service is also referred to as a public network access technology.

In 1988 the Government selected four operators to run nationwide telepoint networks in Britain. The companies were:

—BYPS, originally a consortium of companies but now wholly owned by Hutchison Telecommunications (UK) and called Hutchison Personal Communications;

—Ferranti Creditphone, operating the Zonephone network;

—Mercury Callpoint; and

—Phonepoint, a BT subsidiary.

Of the four operators originally licensed by the Department of Trade and Industry, three—Mercury Callpoint, Ferranti Creditphone and Phonepoint—suspended their services during 1991 due to very slow take-up of services. Thereafter, Hutchison was the only licensee operating a full nationwide telepoint service, marketed under the 'Rabbit' brand name. However, the closure of this service was announced in November 1993.

Personal Communications Networks

Britain will be the first country in the world to offer PCN services. These services are intended to allow the same telephone to be used at home, at work and as a mobile, portable telephone wherever there is network capacity.

PCNs utilise similar technology to existing cellular networks, although the radio cells are smaller and will use GSM digital technology from the outset. The PCN initiative was facilitated by the Government opening up more of the radio spectrum and permitting use of frequencies on the 1.8 GHz range of radio chan-nels. Smaller digital cells at the higher frequencies provide the ben-efits of increasing the capacity of the system to accommodate more users and allowing the use of small lightweight radio telephones. The Government expects PCN to become a mass-market product which will eventually provide a competitive alternative to the fixed telephone network.

In 1989 the Government selected three consortia to provide PCN services—Mercury Personal Communications Network (Mercury PCN), Microtel Communications and Unitel. Since then, Mercury PCN and Unitel have merged to form Mercury Personal Communications (MPC), now trading as Mercury One-2-One, and Hutchison Telecommunications (UK) has acquired Microtel Communications and the new company is known as Hutchison Microtel. Mercury launched a service in the London area in the summer of 1993, and Hutchison will follow in 1994 with a nationwide service.

Mobile Data

Mobile data is one of the latest mobile communications tech-nologies in Britain. It enables two-way data communications between portable, mobile terminals such as fax machines and per-sonal computers, and host computers. In 1989 the Government licensed four companies to provide both mobile and fixed data ser-vices—Cognito Group (now in different ownership and called Cognito Ltd), Hutchison Mobile Data, Paknet and RAM Mobile Data.

Satellite Communications

Satellite communications were first used in Britain in 1962 when the space communications radio station at Goonhilly Downs, Cornwall, took a leading part in the first international experiments in communications using artificial satellites. Since that time there has been a rapid growth in the use of satellite technology for the provision of international telecommunications and television services.

International Satellite Organisations

A substantial proportion of telephone and data traffic to and from Britain is being increasingly carried by satellites. Due to the high cost involved, telecommunications companies are usually unable to provide global satellite communications on an individual basis. Three major consortia have therefore been set up to handle inter-continental satellite communications:

—the International Telecommunications Satellite Organisation (INTELSAT);

—the International Maritime Satellite Organisation (INMARSAT); and

—the European Telecommunications Satellite Organisation (EUTELSAT).

The majority of investors in these three organisations are international network operators.

Both BT and Cable & Wireless are shareholders of the world's largest satellite organisation, INTELSAT, with investments in the consortium of 8 per cent and 3 per cent respectively. INTELSAT has 127 member countries and also users in a number of non-member countries. Founded in 1964, the organisation operates a network of 20 satellites in a geostationary orbit 36,000 km (22,500 miles) above the equator which relay telephone calls to different parts of the earth's surface. It also leases transponders on the satellites to telecommunications operators in individual countries.

Britain is the second largest shareholder in INMARSAT, with an 11 per cent holding. With signatories in more than 70 countries, INMARSAT was set up in 1978 to address the communications needs of those travellers who were beyond the reach of other communications means, such as mariners and aircraft pilots. Services available via INMARSAT include direct-dial telephone, telex, facsimile and high-speed data transmission using 11 satellites and over 26,000 earth stations mounted on ships, oil rigs and other vessels.

In-flight operator-controlled telephone call facilities are available via Portishead radio station near Bristol (in Avon), the busiest coastal radio station in the world. The Aerial 7 satellite dish at Goonhilly is used for BT's Skyphone, a telephone service to and from aircraft on transatlantic routes using specially-equipped British Airways aircraft. The first trials of Skyphone were made in 1988, and a commercial direct-dial telephone service was launched in 1990.

Britain also has one of the largest holdings, at 18 per cent, in EUTELSAT, a Paris-based organisation with 36 members. EUTELSAT was established in 1977 to manage a European communications satellite system. Its main business is television distribution, but the organisation also offers a range of services, including direct-dial telephony services and high-speed data transmission.

The linkage of BT and Mercury to international satellite systems is provided by a number of earth stations located throughout Britain. BT operates satellite earth stations at Goonhilly Downs, Madley (near Hereford), London Docklands and Aberdeen. Aerial 6, Goonhilly's largest dish, entered service in 1985 to provide further capacity on the transatlantic route. The main business systems terminal in London Docklands, with its three 13-metre dish aerials and five smaller aerials, handles cable television channels such as Sky Channel, transmitting them throughout Britain and Europe. Using the latest digital techniques, advanced business data and videoconferencing services to Europe and North America are also carried. The Aberdeen earth station was opened in 1986 at Bridge of Don to meet the specialised needs of the offshore oil and gas industry.

Mercury has three satellite communications centres, in London Docklands, at Whithill in the Midlands and at Brechin in Tayside. The third satellite earth station was opened by the company in 1991 and is providing communications for the offshore oil industry and satellite links between Scotland and the mainland of Europe.

Development of Specialised Satellite Services

In recent years there has been strong growth in the development of specialised satellite services in Britain, stimulated by the licensing of additional satellite service operators, progressive deregulation of the market at the end of the 1980s and the rapid development of satellite technology.

One of the first steps in liberalising satellite communications in Britain was taken in 1984. Prior to that date the provision of telecommunications services via satellite was the sole responsibility of BT. Under the Telecommunications Act 1984, the BT and Mercury duopoly in fixed network service provision was established, but certain areas formed an exception to the general duopoly policy, including the provision of specialised services by satellite. A government statement at the time of the Act also made clear that policy in this area would be kept under review.

In 1988 seven specialised satellite service operators (SSSOs) were selected to provide the full range of one-way point to multipoint satellite services, whether voice, vision or data, for reception initially in Britain. In 1989, the seven operators were permitted to provide such services for reception also in Europe. A range of services was developed by the SSSOs in Britain, including data services for the financial sector, business television and the broadcast of horse-racing and other sporting events to betting offices. The seven SSSOs were:

—British Aerospace Communications;

—BSkyB, the operator of the British satellite channel, Sky;

—Electronic Data Systems (EDS), the US information technology company;

—Kingston Communications;

—Maxwell Communications, which is owned by the French telecommunications company, France Cables et Radio;

—Satellite Information Services (SIS), a British communications company; and

—Uplink, a subsidiary of the communications company Reuters.

In addition to BT, Mercury and the seven SSSOs, the BBC was permitted to operate satellite services for news gathering and transmission purposes.

Also in 1988, the use of receive-only satellites was further liberalised by the issue of a class licence, allowing anyone in Britain to receive voice, data or vision signals from a satellite, of whatever type and wherever they originate. The signal received cannot be passed on beyond the premises on which the receiving satellite dish is located.

In the 1991 duopoly review, the terms of the SSSO licences were re-examined, and as a result a new class licence was introduced for satellite service networks. This enables any company to build and operate a full range of two-way satellite services, including both domestic and international services. The class licence covers one-way point to multi-point and two-way services, although connection of the satellite terminals to the public switched network is not permitted except in limited circumstances. The SSSOs' licences have now been subsumed by the satellite class licence. All licensees must also make arrangements to avoid radio interference.

In February 1992 the Government announced that it had agreed in principle to license Alpha Lyracom and British Aerospace Communications to provide international satellite services that went beyond what is permitted in the class licences. The individual licences would allow voice services to be provided, with interconnection to the public switched network at one end of a call, and data services without any restrictions on connection to the public switched network. The announcement also welcomed other applications from companies wishing to provide similar services. In April 1993 the first three licences were issued to PanAmSat (formerly Alpha Lyracom), E-Sat Telecommunications and Satellite Information Services.

The liberalised market for satellite communications in Britain and technological advances are encouraging the development and use of very small aperture terminal (VSAT) technology. These are two-way services from a satellite network based on a single large hub earth station communicating with a number of small transmit-and-receive earth stations referred to as VSATs.

VSATs are widely used in the United States and are being increasingly used in Eastern Europe as an interim communications technology while operators construct terrestrial networks. In Britain, BT has launched a VSAT service, with its hub in London.

In May 1993 Britain signed a memorandum of understanding on satellite licensing with France, Germany and the Netherlands. This expedites the process of obtaining satellite licences in all four countries, and will help British companies that want to set up VSAT networks in Europe to get the appropriate licences.

International Communications

The number of outbound international telecommunications calls has increased by more than 60 per cent over the period 1988–92. International telecommunications traffic to and from Britain is carried by terrestrial and submarine cable systems and high orbit satellites. In 1990, over 1,000 million outbound calls were made from Britain, which represents around 3 per cent of all the calls made in the country.

Submarine and Terrestrial Cabling Systems

The main part of the international telecommunications traffic to and from Britain is transmitted by terrestrial and submarine optical fibre cables. BT, Mercury, and Cable & Wireless have all contributed to the development of international optical fibre cabling systems, and particularly to the installation of cables between Britain and the United States.

The first transatlantic telephone cable (TAT 1) was laid between Scotland and Newfoundland in 1956, and by 1983 the seventh cable, TAT 7, had been inaugurated between Europe and the United States and Canada. TAT 7 can carry up to 4,200 telephone calls simultaneously. Optical fibre technology has since enabled further improvements to international cabling systems. The world's first international submarine optical fibre cable provided on a commercial basis was laid between Britain and Belgium. Inaugurated in 1986 and costing some £10 million, the cable can carry around 12,000 telephone calls simultaneously.

TAT 8, the first transatlantic optical fibre cable and the first to incorporate an undersea junction, entered service in 1988. It extends from Tuckerton, New Jersey, in the United States to a point west of the French coast, from where one branch runs 800 km (500 miles) to Cornwall and the other 300 km (190 miles) to Brittany in north-west France. BT has contributed the second largest share in the project, which has a capacity of 40,000 simultaneous telephone calls and provides high-speed data transmission and other digital services.

A further high capacity project, TAT 9, completed in 1991, is the result of an agreement between BT, the French and Spanish telecommunications operators, France Telecom and Telefonica, the American Telegraph and Telephone Company (AT&T), and the Canadian telephone operator, Teleglobe. Costing some £250 million, this submarine transatlantic cable is 9,000 km (5,600 miles) long and can carry the equivalent of 80,000 simultaneous voice calls. The transmission rate of 565 megabits per second is double that planned for earlier transatlantic cables. TAT 9 will be used to keep pace with the increasing flow of fax messages and data communications, as well as telephone calls, between Europe and North America.

In 1988 the first optical fibre cable between Britain and France, linking Brighton and Dieppe on France's northern coast, a distance of 150 km (95 miles), was installed after an agreement between BT, Mercury and France Telecom. At the time, 'UK-France 3' was the largest international optical fibre link not requiring repeaters. It carries the equivalent of up to 11,000 telephone conversations at the same time. Two similar links, UK-France 4 and 5, have since come into operation between the two countries. Likewise, other cable systems of the same type, between Britain and the Netherlands, and Britain and Germany, have been set up by BT, Mercury and the telecommunications operators of the Netherlands and Germany.

The latest European submarine fibre optic cable was brought into operation in 1991. 'UK-Germany 5' is currently the world's highest capacity submarine cable, designed to be capable of carrying approximately 200,000 simultaneous calls.

Through the second half of the 1980s, Cable & Wireless pursued a strategy of linking customers in the world's major business centres by its Global Digital Highway (GDH). The GDH is a series of advanced digital networks in countries such as Britain, the United States, Hong Kong and the countries of the Caribbean region, linked by fibre optic submarine cables. The main infrastructure for the GDH was completed in 1991 when the North Pacific Cable entered service. It is the largest trans-Pacific telecommunications cable, capable of carrying up to the equivalent of 85,000 simultaneous calls and is a key link between the United States and the rest of the Pacific Rim. The transatlantic section of Cable & Wireless's GDH is the first such cable to be privately funded.

Satellite Systems

While the majority of transatlantic traffic to and from Britain is carried by TAT and PTAT systems, a substantial amount of other intercontinental traffic is handled by high-orbit satellites. For the contribution of British companies to the development of international satellite communications, see pp. 61–6.

International Traffic Flows

The international telecommunications traffic flow from Britain reflects the level of international trade between Britain and other nations. The highest number of outgoing calls from Britain are made to the United States, followed by Germany and then France. Table 6 shows the major destinations of outgoing calls from Britain and the number of calls made.

Table 6: Britain and its Major Telecommunications Correspondents, 1990

Destination	Outgoing MiTT[a] (millions)	Share of Total (%)
United States	530	20.9
Germany[b]	250	9.9
France	230	9.1
Irish Republic	180	7.1
Netherlands	110	4.3
Italy	100	4.0
Spain	100	4.0
Switzerland	90	3.6
Canada	90	3.6
Austria	80	3.2
Belgium	60	2.4
Sweden	60	2.4
Japan	50	2.0
Denmark	40	1.6
Norway	40	1.6
Other	520	20.3
Total	2,530	100.0

Source: International Institute for Communications.
[a] MiTT is minutes of telecommunications traffic. Data are for public telephone voice circuits only. Totals are rounded to the nearest 10 million MiTT.
[b] Data for Germany include outgoing calls made from Britain to the former German Democratic Republic.

People in Britain make among the highest number of calls in Europe. Around 600 million outbound international calls were made in Britain in 1989, compared with 240 million calls in Italy, which has a population of similar size to Britain. Germany placed 800 million outbound international calls and France made 1,682 million minutes' worth of calls. Britain's international tariffs are the cheapest in the EC. Table 7 compares the quantity of international calls originating from specific countries.

Table 7: Total International Telephone Traffic, 1989

	Outgoing traffic (million calls)	Outgoing traffic (million minutes)
United States	1,200	
Germany[a]	800	
France[b]		1,700
Britain	600	
Netherlands	220	
Italy	240	
Belgium	170	
Spain	160	

Sources: AT&T World's Telephones; the International Telecommunications Union.

[a] Data for Germany include outgoing calls made from the former German Democratic Republic.

[b] Statistics for France are based on the length of international calls.

Cable Television

The development of cable television in Britain began at the end of 1983, when the Government awarded 11 pilot franchises for operators to deliver cable television programmes and other telecommunications services over local broadband cable networks. In addition to giving customers greater choice in the provision of entertainment services, cable television companies could provide a full range of telecommunications services, although voice telephony had to be provided in conjunction with BT or Mercury.

Following the passage of the Cable and Broadcasting Act 1984, the Cable Authority was established with the objectives of awarding further cable television franchises and promoting the benefits of broadband cable. A total of 135 cable television franchises were awarded by the Cable Authority, covering nearly 70 per cent of the British population. Of the 63 cable franchise areas currently in operation, 39 are providing full telephone services, primarily voice communications. The Cable Authority was subsequently wound up under the Broadcasting Act 1990, and its function of awarding cable licences has passed to the Independent Television Commission set up under the same Act of Parliament.

Prior to the 1990 duopoly review, cable television companies could only offer voice services in their franchise areas if they had reached a commercial, or so-called 'interconnection', agreement with either BT or Mercury or both, covering the service charges, terms and conditions. The operation and maintenance of the equipment used to provide these services was the responsibility of

BT or Mercury, and the cable companies were unable to run such network equipment in their own right.

However, following the White Paper (see p. 17), the Government decided that, in the interests of promoting further competition in the provision of local telecommunications services, the restrictions covering the interconnection of cable television networks with BT and Mercury networks should be removed. Since that date, a number of cable operators have bought or leased their own independently controlled telecommunications switching systems, and Cambridge Cable and South Herts Cable, in Watford, have also installed their own switches. Several other operators are in the process of obtaining switches for other franchises. A wide range of interconnection agreements have been negotiated between the cable television operators and BT and Mercury.

The cable television operators providing telecommunications services at the end of 1992 were:

—Birmingham Cable;

—Cable London;

—Cable Telecom in Windsor and Slough;

—Cambridge Cable;

—Diamond Cable in Nottingham;

—Encom Cable TV and Telecommunications in London;

—General Cable in various franchises throughout Britain;

—Jones Cable in south Hertfordshire;

—Nynex Cablecoms in the south of England;

—Southwestern Bell in the Midlands;

—United Artists Communications in the south of England and Edinburgh; and

—Videotron Corporation in various franchises throughout Britain.

A direct broadband cable television line into the home allows customers, by payment of a subscription charge, to receive both cable television channels and basic telephony services. Specialised equipment, such as a telephone with an access button, is not required.

Telephony is the fastest growing element of the cable business in Britain. There was a fivefold increase in the number of telephone lines installed by cable television operators during 1992, increasing from more than 21,000 at the end of 1991 to about 107,000 one year later. By October 1993, 245,000 telephone lines were connected.

Equipment Manufacture

The British telecommunications equipment industry is one of the longest established in the world. It has a turnover of about £2,500 million and exports some £620 million worth of equipment to over 40 countries. It spends more than £500 million on research and development each year, and a number of the companies concerned have received the Queen's Award for Technological Achievement.

In addition to GPT (see below) and a variety of smaller British firms, overseas companies have also invested in telecommunications equipment manufacturing in Britain. During the 1980s government deregulation of the supply of telecommunications apparatus in Britain strongly stimulated demand for equipment. Since 1981 private sector organisations have been able to supply, install and maintain most types of subscribers' equipment attached to the public telephone network.

GPT

GPT Limited (formerly known as GEC Plessey Telecommunications) is Britain's foremost telecommunications equipment manufacturer, producing a wide range of equipment. The company operates worldwide, with subsidiaries and associated companies in 21 countries, including the United States, Hong Kong and Russia. It employs around 12,000 people, invests over £150 million a year in research and development, and in 1992–93 had a turnover of £1,125 million with a pre-tax operating profit of £127 million.

GPT is a world leader in digital public switching, with its System X range, and a European leader in digital private switching. It is also a leading supplier of 'intelligent' payphones, with 59 customers in 46 countries, and of videoconferencing equipment, with 1,000 installations in 40 countries. It has supplied videotex systems to more than 80 per cent of the world's public videotex networks and telecommunications systems to 113 countries.

Other Manufacturers and Technologies

In addition to GPT, there is a variety of companies in Britain which have considerable strengths in technologies vital to advanced telecommunications systems.

Optoelectronics and Cables

The use of glass fibres to carry information by optical pulses was first demonstrated in Britain in laboratories which now form the main research centre of Northern Telecom Europe Ltd. Since then, British companies manufacturing optical fibre cables and optoelectronic components have acquired prominence in world markets.

BT & D Technologies is a leading manufacturer of optoelectronic components for use with fibre optic communications systems, including cable television. The company's exports, which make up 80 per cent of sales, are currently about £20 million a year.

The Plessey Research Centre at Casswell and the optoelectronics division of Northern Telecom in Paignton are recognised as centres of excellence. BICC Cables, which claims to be the second largest cable manufacturer in the world after Alcatel, has substantial facilities for the production of optical fibres and associated components, while Fulcrum Fujitsu is a leader in the development of optical fibre networks. Nearly half of the world's submarine cables—some 240,000 km (150,000 miles)—have been supplied by STC Submarine Systems, a major manufacturer of submerged telecommunications systems.

Cordless Communications

Britain has pioneered the development of a standard for digital cordless telephony (CT2/CAI), which has been adopted by a number of countries in Europe and is attracting much interest from many other companies around the world. Shaye Communications is developing a range of innovative products for this growing market and is backed by the resources of AT & T.

The research and technical skills available in Britain have encouraged several major foreign companies to set up manufacturing facilities. Foremost among these is Motorola, which manufactures analogue and GSM digital mobile telephones in Scotland. The plant is Motorola's largest handset manufacturing facility in Europe, and when its present expansion is completed will employ over 2,000 people and represent an investment of over £100 million. Its manufacturing facility in Swindon was recently expanded to make it Motorola's main centre in Europe for mobile telephony cellular infrastructure equipment.

High Speed Transmission

Companies such as GPT, BNR Europe Ltd and Netcomm are in the forefront of the development of a new generation of switches and network products which will enable networks to transmit information at high speeds and deal with rapidly increasing network traffic.

Compression Technologies

Several British companies have made significant advances in the development of sophisticated data compression technologies which will facilitate the simultaneous transmission of voice traffic, data and moving images. The availability of such technologies has led to the development of videoconferencing products, in which GPT is a world leader. Other leading British companies in this field include BT, National Transcommunications and Snell and Willcox.

Telecommunications Software

Telecommunications is becoming increasingly dependent on software as the number, complexity and variety of services proliferate. Britain's manufacturing capability is complemented by the high quality of software supplied by a number of companies which have a long-established reputation in world markets. Foremost among these are Logica, IPL, GPT Syntegra, ICL and Kingston-SCL. Their collective strengths meet needs for applications such as mobile telecommunications, voice processing, billing and customer care systems, multimedia and intelligent network management. Multinationals such as Ericsson, Motorola, IBM and Hewlett Packard have significant software facilities in Britain.

Addresses

Department of Trade & Industry, Telecommunications and Posts Division, 151 Buckingham Palace Road, London SW1W 9SS.

BBC Research & Development, Kingswood Warren, Tadworth, Surrey KT20 6NP.

BICC Cables Ltd, Helsby, Warrington WA6 0DJ.

British Approvals Board for Telecommunications, Claremont House, 34 Molesey Road, Horsham, Walton-on-Thames KT12 4AQ.

British Telecommunications plc, 81 Newgate Street, London EC1A 7AJ.

BT & D Technologies, Whitehouse Road, Ipswich, Suffolk IP1 5PB.

Cable and Wireless plc, 124 Theobalds Road, London WC1X 8RX.

Fujitsu Fulcrum Communications Ltd, 113 Fordrough Lane, Birmingham B9 5LD.

GEC-Marconi Materials Ltd, Caswell, Towcester, Northants NN12 8EQ.

GPT Ltd, New Century Park, PO Box 53, Coventry CV3 1HJ.

Hewlett Packard Ltd, King Street Lane, Winnersh, Wokingham, Berks RG11 5AR.

IBM UK Ltd, National Enquiry Centre, 389 Chiswick High Road, London W4 4AL.

ICL (UK) Ltd, Government & Major Companies Division, Observatory House, Windsor Road, Slough SL1 2EY.

Independent Television Commission, 70 Brompton Road, London SW3 1EY.

INMARSAT, 40 Melton Street, London NW1 2EQ.

IPL, Eveleigh House, Grove Street, Bath BA1 5LP.

International Telecommunications Union, Place des Nations, 1211 Geneva 20, Switzerland.

Kingston-SCL Ltd, Clerwood House, 96 Clermiston Road, Edinburgh EH12 6UP.

Logica Space and Communications Systems Ltd, 69 Newman Street, London W1A 4SE.

Mercury Communications Ltd, New Mercury House, 26 Red Lion Square, London WC1R 4HQ.

National Telecommunications Ltd, Crawley Court, Winchester, Hants SO21 2QA.

Netcom Ltd, 3 Olympic Business Centre, Paycocke Road, Basildon, Essex SS14 3EX.

Northern Telecom Europe Ltd, Radio Infrastructure, Brixham Road, Paignton, Devon TQ4 7BE.

OFTEL, Export House, 50 Ludgate Hill, London EC4M 7JJ.

Shaye Communications Ltd, Capital House, 48–52 Andover House, Winchester, Hants SO23 7BH.

Snell and Wilcox, 57 Jubilee Road, Waterlooville, Hants PO7 7RF.

STC Submarine Systems Ltd, Christchurch Way, Greenwich, London SE10 0AG.

Syntegra, Guidion House, Harvest Crescent, Ancells Park, Fleet, Hants GU13 8UZ.

Telecom Securicor Cellular Radio Ltd (Cellnet), 1 Brunel Way, Slough SL1 1XL.

Vodafone Group, The Courtyard, 2–4 London Road, Newbury, Berkshire RG13 1JL.

Further Reading

£

Competition and Choice: Telecommunications Policy for the 1990s. Department of Trade and Industry. ISBN 0 10 113032 5.	HMSO 1990	8.80
Performance Indicators for Public Telecommunications Operators. OECD. ISBN 92 64 13403 4.	HMSO 1990	24.00
Universal Service and Rate Restructuring in Telecommunications. OECD. ISBN 92 64 13497 2.	HMSO 1991	24.00
Yearbook of Common Carrier Telecommunications Statistics (20th edition). ITU. ISBN 92 71 18061 5.	ITU 1993	[a]
A Consumer Guide to Telephones and Services.	OFTEL	Free
Guide to Cellular Radio.	OFTEL	Free
The Regulation of BT's Prices.	OFTEL	Free
Future Controls on BT's Prices.	OFTEL	Free
Telephone Services in 1991.	OFTEL	Free

Annual Reports
British Telecommunications plc.
Cable and Wireless plc.
OFTEL.
Vodafone Group.

[a] Price 68 Swiss francs

Abbreviations and Glossary

ACD	automatic call distribution
AXE 10	a digital telephone exchange switch for the public network manufactured by Ericsson of Sweden
BABT	British Approvals Board for Telecommunications
Broadband	refers to transmission channels which have a bandwidth greater than that of standard telephone voice channels
BTMC	BT Mobile Communications
CCITT	International Telecommunications and Telegraph Consultative Committee
Centrex	enables all PBX facilities (such as inbound directly dialled calls and call forwarding) from specialised equipment located at public exchanges switch, so users do not have to purchase PBX equipment
CEPT	Conference of European Posts and Telecommunications Administrations
Class licence	a licence granted by the Department of Trade and Industry to a particular system class or company
CT2	second generation cordless telephone digital technology, whereby a CT2 cordless telephone uses a radio link between the handset and a public CT2 (or telepoint) base station
DTI	Department of Trade and Industry
Duopoly	the regulatory policy under which only BT and Mercury were permitted to operate fixed telecommunications services in Britain
EDI	electronic data interchange, the interchange of business information and transactions using electronic messaging systems
E-Mail	electronic mail, the electronic transmission and reception of text-based information and messages without the need for the recipient to be present at the time of transmission. Public E-Mail messages are transmitted over PSDNs

Equal access	allows customers of a local network (the 'access' network) to send their long distance calls over the network of their chosen trunk operator
ERMES	European Radio Messaging System, a pan-European radiopaging standard
ETACS	extended total access communications system
ETSI	European Technical Standards Institute
Euromessage	a proprietary pan-European radiopaging service provided by certain European operators, and an interim service before ERMES services are fully available throughout the EC
EUTELSAT	European Telecommunications Satellite Organisation
GDH	Global Digital Highway
GNS	Global Network Services
GSM	Groupe Speciale Mobile of CEPT which defined the European standard for digital cellular communications
GVPN	global virtual private network
ICSTIS	Independent Committee for the Supervision of Standards of Telephone Information Services
IIC	International Institute of Communications
INMARSAT	International Maritime Satellite Organisation
INTELSAT	International Telecommunications Satellite Organisation
ISDN	integrated services digital network, a multifunction, fully integrated, digital telephone system, capable of handling voice, data, text and image transmission
ISR	international simple resale—the use of international leased circuits connected to the public switched network at both ends
ITC	Independent Television Commission
LAN	local area network
MCN	Micro Cellular Network, a proprietary network of Vodafone based on GSM technology
MMC	Monopolies and Mergers Commission
M-VTS	Marconi videotelephone standard
OECD	Organisation for Economic Co-operation and Development

OFTEL	Office of Telecommunications
ONP	open network provision
PBX	a private branch exchange that handles incoming and outgoing calls from a customer's premises unattended and is linked to the public telephone network
PCN	personal communication network, which enables two-way communications using GSM technology
PMR	public-access mobile radio, a two-way radio-based mobile communications technology using specifically allocated radio frequency spectrum
PRS	Premium Rate Service
PSDN	packet switched data network, which provides dedicated data transmission services whereby individual packets of data of a set size and format are accepted by the network and delivered to their destination; the CCITT standard for PSDNs is known as X.25, hence PSDNs are often referred to as X.25 networks
PSTN	public switched telephone network
PTAT	private transatlantic telephone cable
PTO	public telecommunications operator
Radiopaging	one-way radio-based communications technology
RPI	Retail Prices Index
Simple resale	the resale of network capacity to a third party where there are one or two points of interconnection between a customer's private network and the public network
SSSO	specialised satellite services operator
System X	a digital telephone exchange switch for the public network manufactured by GPT
TACS	Total Access Communications System
TAT	transatlantic telephone cable
Telepoint	based on second generation CT2 digital technology, whereby a radio link is used between a telephone handset and public telepoint base station. Light, portable telephones are used to provide one-way communications
Telepoint-to-the-home	a public base station is replaced by a small private base station used in the home or office

Telex	One of the earliest forms of transmitting text messages based on analogue technology
TSL	telecommunications services licence
VADS	value-added and data services
VSAT	very small aperture terminal

Index

Printed in the UK for HMSO.
Dd 0297146, 2/94, C30, 56–6734, 5673

TITLES IN THE ASPECTS OF BRITAIN SERIES

FORTHCOMING TITLES